Fruits and Vegetables of the Caribbean

M.J. Bourne, G.W. Lennox, S.A. Seddon

CARIBBEAN

Macmillan Education
Between Towns Road, Oxford OX4 3PP
A division of Macmillan Publishers Limited
Companies and representatives throughout the world

www.macmillan-caribbean.com

ISBN 0 333 45311 5

Text © M.J. Bourne, G.W. Lennox & S.A. Seddon 1988
Design and illustration © Macmillan Publishers Limited 1988

First published 1988

Printed and bound in Thailand

2009 2008 2007 2006 2005
24 23 22 21 20 19 18 17 16

Acknowledgements

The authors wish to thank the many people in the Caribbean for their generous
assistance in researching information and in locating suitable specimens for
photography.
Particular thanks are due to:
Iris Bannochie, Andromeda Gardens, Barbados
Omar and Flo Delfosse, Jamaica
Kay Donawa, Information Officer, Barbados Agricultural Development Corporation
Richard Humphrey, Undercover Vegetables, Trinidad
James Joseph, St Lucia
Earl Kirby, St Vincent
John Lewis, Dominica
Alphonso Mclean, Terra Nova Hotel, Jamaica
Jualla Maharaj, Sutton Farms, Trinidad
Donald Menzie, Andon Farm, Jamaica
George Money, Barbados
Eric Patience, Asa Wright Nature Centre, Trinidad
Tai and Rob Stewart, Trinidad
Ivan Waterman, Barbados

Contents

Photographic Acknowledgements
The authors and publishers wish to acknowledge, with thanks, the following
photographic sources.
John and Iris Bannochie pp. 8; 27 right
Anne Bolt pp. 10; 55
D.Y. Mordecai p. 50
Rex Parry p. 52
Eddie Sinclair p. 31
All other photographs were taken by the authors.
The publishers have made every effort to trace the copyright holders, but if they
have inadvertently overlooked any, they will be pleased to make the necessary
arrangements at the first opportunity.

Introduction

When Christopher Columbus came to the Caribbean at the end of the fifteenth century, he thought he had found the Garden of Eden. His diaries record the richness of the vegetation and the great variety of strange and colourful fruits he found there. The chain made up by the Caribbean Islands stretches 2500 miles from the coast of Florida to the northern edge of Venezuela. In all, this chain comprises more than 7000 islands. Some are nothing more than small rocks in the Caribbean Sea, but others such as Cuba, Hispaniola and Jamaica are large islands with an immense variation in topography and climatic conditions. It is this variety of habitats with their plentiful supply of rain and warmth which allows the islands of the Caribbean to grow such an enormous range of exotic fruits and vegetables.

There is a good deal of confusion about the terms 'fruit' and 'vegetable'. Many people think of a fruit as something which is soft, succulent and juicy, but this is not always the case. The word 'fruit' has a strict botanical meaning. It is a part of a plant which develops from the female parts of the flowers after pollination and fertilisation have taken place. So, for example, a lime is a fruit, so is a pepper, a coffee berry and a christophine. The word 'vegetable' is used much more generally. Some things which we loosely call vegetables are, botanically speaking, fruits. An egg plant is a fruit and so is a pumpkin. Other parts of the plant we eat are not derived from the flower. The sweet potato, yam and dasheen are root tubers. They are also called vegetables. Even the leaves of the dasheen are called vegetables when used to make delicious Caribbean soups.

It is beyond the scope of this small book to describe all the fruits and vegetables found in the various Caribbean islands. Many, such as the avocado, cassava and the cocoa tree were known to the Arawak Indians long before the arrival of Columbus. Others, such as the breadfruit, the akee and the banana have been introduced to the Caribbean from other parts of the tropics. The fruits and vegetables described in this book comprise the main types which the average visitor or tourist will see on roadside stalls, in local markets and at dining tables in hotels throughout the region. Some of the entries have been selected for reasons in addition to their commonality and use in different islands. Particular species are of great historical interest. Others are a culinary delight and need to be known to the gourmet and all those interested in Caribbean cooking. The reader will note the absence of some very common fruits and vegetables. These have been omitted because of space limitation and also because the visitor to these lovely islands will probably be interested in the more exotic forms not so well known in his own country.

In recent years there has been a migration of people, fruits and vegetables from the Caribbean to other parts of the world. The inhabitants of North America and Europe are more familiar with fruits such as aubergines, plantains and avocados than they were twenty years ago and this trend is likely to continue. Even so, the reader must remember that some fruits are poor travellers and themselves need to be visited. In any case, the finest specimens of fruits and vegetables are probably not found in the shops and supermarkets of North America and Europe. They are seen at their best in their natural surroundings, enjoying the wonderful Caribbean sunshine. It's up to you to visit them.

We hope you enjoy reading about the fruits and vegetables we have chosen to present to you. We would certainly like to think that you will be sufficiently stimulated, having read this small book, to go and search for other interesting types during your stay in the islands. There are plenty to chose from and you will have no difficulty in finding new varieties to try. Your job will be infinitely easier than ours has been in selecting from the enormous number available, the forty-eight fruits and vegetables included in this book.

Family Sapindaceae

Akee (*Blighia sapida*)
Other names Ackee, Akee Akee, Vegetable Brain, Achee

The akee tree was brought to the Caribbean during the slave trade and may well have been carried by Captain Bligh himself on board the infamous HMS *Bounty* in 1787. The tree is of medium size with compound leaves and small, greenish-white flowers which have a pleasant smell. The fruits develop thick, reddish-orange skins which enclose shiny black seeds, each with a fleshy, whitish-coloured structure at its base called an aril. The aril is the edible part. However beware! The fruit is ready to eat only when it has turned red and has split open. Unripened arils and those overripe are poisonous and 'Jamaica Poisoning' is the term given to the condition resulting in death caused by eating arils at the wrong stage of development. Jamaica is the only island in the Caribbean where akee is eaten in any quantity. Akee and salt fish is a favourite Jamaican dish and despite the fact that the cod is soaked overnight to reduce its saltiness, it still maintains enough of a 'salty' taste to combine with the mildness of the akee to produce a delicious 'marriage' for the taste buds.

Family Marantaceae

Arrowroot (*Maranta arundinacea*)

Other names West Indian Arrowroot, St Vincent Arrowroot, Bermuda Arrowroot

Arrowroot is the name given to the starch products of a number of different plants. However, true arrowroot is the starch obtained from the rhizome or underground stem of an herbaceous perennial plant called *Maranta arundinacea*. This plant grows wild in the northern part of South and Central America. It has been introduced to a large number of countries in the tropics but only one cultivates it on a large scale: this is the island of St Vincent. The leaves of the plant are spear-shaped with veins set very close together. The rhizome, from which the starch is extracted, is light grey in colour with what look like overlapping scales. The whole structure looks similar to the individual flowers of the shrimp plant. Early records suggest that crushed rhizomes were used to treat skin wounds caused by poisoned arrows, hence the name now given to the plant. Since its discovery, the starch obtained from the plant has been used as an antidote to other poisons and, more recently, as the basis of an easily digestible food for invalids. The arrowroot starch has even more diverse functions including the starching of clothes, thickening soups, puddings and various sauces, and also as an ingredient in face powders and glues.

3

Family   Solanaceae

Aubergine (*Solanum melongena*)
Other names   Chinese Egg Plant, Egg Fruit, Melongene, Garden Egg, Jew's Apple, Mad Apple, Brinjal

Aubergine is a common plant throughout the Caribbean. It is a member of the potato family and needs a warm climate and a plentiful supply of rain to produce good quality fruits. The plant bears fruit about five months after the initial planting and the fruits are ready for picking when they have swollen and their skins have turned a dark, glossy purple. Aubergines are served as a vegetable either boiled or cut into strips and fried. Sometimes they are stuffed with small pieces of meat and baked in the oven. In Trinidad, they are also used to make fritters called *baigani*. Many visitors will be familiar with the Greek dish moussaka, in which aubergines are an important ingredient. Aubergines have very little nutritional value; their contents are more than ninety per cent water.

Family Lauraceae

Avocado (*Persea americana*)
Other names Alligator Pear, Aguacate, Midshipman's Butter

The avocado is a native of Central America and was certainly in use when Columbus first came to the Caribbean in 1492. The tree on which the avocado grows attains a height of about thirty feet and has dark green, shiny foliage and clusters of small, inconspicuous flowers, which are light green in colour. It is not an easy tree to identify except for its characteristic fruit which, because of its shape, is sometimes called an avocado pear. In fact, avocados vary enormously in shape. The majority are pear-shaped, but others are oval or round. The skin is thick with a warty appearance and the flesh inside surrounds a large, oval-shaped stone. The flesh is light yellow and soft when ripe. Avocados are eaten in salads and as an entrée to a main course, and can be made into a delicious mousse. It may also be cut in half and filled with shrimps and pieces of lobster or other forms of shellfish. In some islands, the leaves are made into a drink similar to tea, which is supposed to have medicinal properties in helping to reduce high blood-pressure.

Family Musaceae

Banana (*Musa* spp.)

Bananas and their close relatives the 'figs' and plantains are some of the most important food crops in the Caribbean and constitute the major export crop of the Windward Islands, reaching 138 000 tonnes in 1984. Although it is often described as a tree, botanically the banana isn't a tree at all but a large herbaceous plant growing to a height of twenty feet or more, depending on the species. The plant originates from the Indo-Malaysian region but it has now found its way to most parts of the tropics. The huge leathery leaves grow in a whorl-like fashion at the top of the 'stem' which is, in fact, made up of many overlapping leaf bases. The flowers grow in clusters and hang down on long stalks. Each flower is covered with a dark purple bract and a mature floral structure is a very colourful and remarkable sight. Flowering and fruiting occur throughout the year, but each stem bears only one batch of fruit, after which it dies. In the 1960's, Panama disease attacked large numbers of banana plantations and new disease-resistant strains had to be developed. Bananas are served at breakfast as a starter, in fruit salads and in numerous forms as desserts. They can be fried, 'flamed', made into pancakes and even cooked as sweet banana omelettes. They can also be roasted in their skins and then opened up and sprinkled with chocolate and lime.

Family　Malpighiaceae

Barbados Cherry (*Malpighia glabra*)
Other names　West Indian Cherry, Cerise, Miocise, Acerola, Sour Cherry

This is a small tree which is often pruned back and used for making ornamental hedges in private gardens. The fruit grows to about the same size as the ordinary cherry. The colour of the fruit varies at different stages in its development starting as a dull orange and then gradually darkening to reddish-brown when mature. Most fruits ripen and are ready for picking from April to July. One interesting feature of the Barbados cherry is its very high ascorbic acid (vitamin C) content. Weight for weight, it is much richer in this specific nutritional aspect than most citrus fruits. It is also reported to be unique in that its vitamin C is destroyed less easily by cooking than is the case with other fruits. The Barbados cherry is seldom eaten raw because its taste is too sharp for comfort. However, it is used in puddings or made into jam, jelly and other preserves.

Family Moraceae

Breadfruit (*Artocarpus communis*)
Other names Breadnut Tree

The breadfruit tree grows to a height of about sixty feet and is a common sight on many Caribbean islands. It originates from the South Pacific and was first brought to the Caribbean by Captain Bligh in 1793. Captain Bligh's deep concern for his breadfruit cuttings at the expense of his crew may have been one of the contributory factors in the famous 'Mutiny on the Bounty'. It was thought that the large fruits would provide food for the slaves brought to the Caribbean to work on the sugar plantations. The St Vincent botanical gardens boast a specimen, which it is claimed, has been grown from a sucker of one of Bligh's original trees. The leaves are large and deeply lobed with a clearly veined pattern on the under-side. It has catkin-like male and female flowers. The mature fruits grow in clusters and are spherical when mature, with a green 'pimply' skin. The breadnut is the seeded form of the species, while the fruits of the breadfruit form seldom contain seeds and are always propagated by cuttings. The mature fruit is soft and fleshy with a yellowish-brown colour. It is rich in carbohydrates and vitamins A, B and C. It is cooked as a vegetable with the main meal but it is also made into breadfruit bread, pie and puddings.

Family Anacardiaceae

Cashew (*Anacardium occidentale*)
Other names Cajugaha, Cashew Apple

The cashew tree probably originates from South America and the West Indies, but it has since been taken to many parts of the tropics. The tree itself is not particularly striking and the visitor may be excused for not spotting it amongst the luxuriant Caribbean vegetation. However, its fruit, the cashew apple, is unique botanically and the casual observer will have little difficulty recognising it. The apple itself is the pedicel or swollen stalk of the flower and it is important because of the nut it produces. In some Caribbean islands the fleshy cashew apple is stewed and eaten as a vegetable but it can also be made into a jelly or fermented into a wine or liqueur. The real fruit is the nut borne on the end of the apple and it is the kernel of this nut that is roasted commercially. The price of the nuts is relatively high since they have to be harvested by hand. The oily liquid in the shell of the nut is poisonous and some people have been affected when extracting the nut from its shell. The shell liquid causes irritation of the skin and, in severe cases, brings about death.

Family Euphorbiaceae

Cassava (*Manihot esculenta*)
Other names Manioc, Brazilian Arrowroot, Tapioca

Cassava is a shrub about seven feet in height with dull green, palmate leaves. Although the shrub itself is relatively inconspicuous, it is important agriculturally because of the large tuberous roots which provide the cassava itself. The roots grow in a cluster just beneath the surface of the soil and they swell into large oval-shaped structures as they fill with stored starch. Mature tubers have a greyish-brown, mottled skin and a white, fibrous interior. Cassava was the staple diet of the early Caribbean Indians and is still an important item in the diet of many West Indians. It can be ground into meal which is then used to make a kind of bread. The juice obtained from grated cassava root is flavoured with cinnamon, cloves and brown sugar to make cassareep, an essential component of a stew known as pepperpot. The tubers do contain a certain amount of prussic acid which must be removed by cooking or pressing, otherwise it can cause death by poisoning. Arawak Indians were known to commit suicide by biting into uncooked cassava rather than face torture from their Spanish invaders.

11

Family   Cucurbitaceae

Christophine (*Sechium edule*)
Other names   Choyote, Squash, Cho-Cho, Chayote

The christophine is a member of the gourd family and it originated in Mexico. It is a pear-shaped fruit with a single large stone in the centre. The skin varies in colour from white to pale yellow or bright green. The surface of the fruit is wrinkled and some specimens have a slightly prickly feel to them. It is a large type of squash whose flesh can be eaten raw or, more commonly, cooked as a vegetable to accompany a main meat dish. When cooked properly, it tastes like a fresh, tender young courgette or marrow. If over-cooked it becomes very mushy because of its high water content. The seed is also edible.

Family Sterculiaceae

Cocoa (*Theobroma cacao*)
Other names Chocolate Tree, Cacao

Although the cocoa tree originated in Central and South America it is now found in many parts of the tropics, particularly West Africa where it is grown extensively as a commercial crop. The tree was grown by the Aztecs long before European explorers visited tropical America, and Christopher Columbus took cocoa beans back to the Spanish court in 1502. In 1625, Spanish explorers introduced the tree to Trinidad and large plantations began to be planted there. Today, visitors will also see cocoa trees in islands such as Tobago and Grenada. Although the cocoa tree has been cultivated for the last four hundred years, it wasn't until the middle of the last century that the beans were used to produce chocolate as a confectionery. The tree is easily recognised by its handsome dark green, shiny leaves and by its characteristic pods, each about ten inches long which are often seen growing directly from the trunk. The pods turn from green to brownish-red when ripe and at this stage they are picked. The beans are extracted from the pods and laid out to dry in the sun before being roasted and then used to make commercial cocoa, chocolate and cocoa butter.

Family Palmae

Coconut (*Cocos nucifera*)
Other names Coco Palm, Coconut Tree

Most people's idea of a tropical island includes white, sandy beaches fringed with coconut palms. The islands of the Caribbean certainly live up to this dream; they have coconut palms in abundance. The origin of these tall, graceful palms is lost in history. What is certain is that they are now spread widely through the tropics and they are one of the first sights to greet visitors arriving in the Caribbean. The fruit of the coconut consists of a green or yellowish-brown husk inside which the coconut itself is found. The fibrous husk is waterproof and the protection it gives the nut has resulted in the seeds floating round the world to establish the palms on most tropical beaches. The nut itself has a hard outer shell with a kernel inside. In young 'green' coconuts the kernel is soft and jelly-like and the central cavity is filled with coconut 'milk'. In most islands the visitor will see street vendors offering fresh coconuts to quench the thirst. As the coconut matures, the kernel becomes harder and the amount of milk decreases. The dried kernel is used in all kinds of cooking and helps make items like Barbados' coconut custard pie, Puerto Rico's *tembleque* and Jamaica's coconut tart. Dried coconut is the copra of commerce and is exported. Another important by-product, especially in Dominica, is coconut oil, used in the manufacture of soap. Even the leaves are not wasted and in many islands they are still used for thatching.

15

Family Rubiaceae

Coffee (*Coffea arabica*)

Coffee trees grow in a number of Caribbean islands, including Trinidad and Tobago and Jamaica. The trees are usually between ten and fifteen feet high when mature and they have dark green, glossy leaves. The flowers are small and very fragrant, each with five white petals. The fruits grow in clusters of berries which turn from green to red when mature. Coffee trees require fairly cool temperatures and a moderate rainfall and they also need extra shade. Jamaica grows the famous 'Blue Mountain' coffee in the hills of Surrey County in the eastern part of the island. It is considered by many coffee connoisseurs to be one of the world's finest coffees and much of the crop is exported. Coffee's best-known use is as a beverage, but it is also used in flavouring sweets and cakes. Jamaica's famous liqueur, Tia Maria, is flavoured with Blue Mountain coffee.

Family   Annonaceae

Custard Apple (*Annona reticulata*)
Other names   Sugar Apple, Bullock's Heart

The custard apple belongs to an important group of tropical American small trees and shrubs which are cultivated widely and which produce edible fruits. The tree grows to a height of twenty feet and it has a spreading habit with bushy branches and a greyish coloured bark. The flowers are olive-green or yellow and they grow in clusters in the axils of the leaves. Each fruit is a yellowish-green, heart-shaped structure and has a white, sweet, granular pulp with a custard-like consistency from which the fruit derives its name. Large numbers of glossy seeds are found within the pulp. Mature fruits become reddish-brown in colour and these are eaten raw as a dessert.

Family Araceae

Dasheen and Eddoe (*Colocasia esculenta*)
Other names Cocoyam, Taro, Eddo

The dasheen (pictured left) is an herbaceous perennial with large, handsome leaves, long-cultivated throughout the tropics. The plant grows to a height of about four or five feet. It is cultivated for its underground tubers, which are rich in starch. The tubers are similar in size to small or medium-sized potatoes, usually with a fibrous skin. In Trinidad the plant is grown for both its tubers and leaves. Young leaves are particularly tender and they make a delicious vegetable soup known as callaloo. Most varieties of dasheen are characterised by an acrid taste, but this is usually removed by boiling and cooking. The tubers are often compared to artichokes and some people prefer them to both artichokes and potatoes. The closely-related eddoe (var. *antiquorum*) is often confused with the dasheen (var. *esculenta*). However, the dasheen has a large, central corm (the edible portion) and relatively few side tubers, whereas the eddoe (pictured right) has a rather small central corm, which is not normally eaten, and many small, edible, side tubers. These tubers are usually boiled, roasted or baked; the leaves are not usually consumed.

19

Family Zingiberaceae

Ginger (*Zingiber officinale*)

This is a perennial herbaceous plant, originating in south-east Asia, but now widely cultivated in most tropical countries and particularly common in the Caribbean. Jamaica is an island famous for the production and exportation of commercial ginger and, for this reason, is sometimes called the 'Land of Ginger'. The plant grows as a group of leafy shoots which reach a height of between two and three feet. The ginger familiar in shops and supermarkets is extracted from the underground tuberous stems. The white form of ginger is obtained by washing, boiling, peeling and then blanching the rhizomes. Another form, black ginger, is produced by washing and boiling the rhizomes and then drying them. The dried rhizomes are also ground into a powder. Ground ginger is used mainly in cake flavourings and for adding to drinks such as ginger beer and wine. Preserved ginger, which is used in baking fruit cakes and as an ingredient of ginger marmalade, is produced by boiling prepared rhizomes in a sugar syrup. Ginger is also the source of an important oil called gingerol.

Family Rutaceae

Grapefruit (*Citrus paradisi*)
Other names Pomelo

The grapefruit was not amongst the rich array of exotic fruits which greeted the early explorers from Europe when they came to the Caribbean in the fifteenth century. Nor was it brought to the islands from other parts of the tropics by some enterprising sea captain. It simply didn't exist anywhere before the eighteenth century. It is a Caribbean creation! A British sea captain named Shaddock sailed to Jamaica from the Pacific Islands of Polynesia about 1750. Amongst his cargo he carried some citrus fruits with very thick skins and a reddish-yellow flesh, which had an aromatic but bitter taste. The fruits were soon planted and because they found the soil and climate to their liking, they grew quickly and soon spread to other islands. Over the years, either through natural selection or as a result of hybridisation between the shaddock and a sweet orange, a new type of citrus fruit appeared. This was the grapefruit. Today, the grapefruit is a favourite breakfast starter in many countries and visitors to the Caribbean will find it on many hotel menus. The total world annual production of grapefruits is in excess of four million tonnes — a measure of this remarkable fruit's popularity.

Family Leguminosae

Groundnut (*Arachis hypogaea*)
Other names Peanut, Monkeynut, Earth Nut, Goober-pea

This small, annual, trailing plant is native to Brazil but is now cultivated throughout the tropics. A unique feature of the plant is its habit of producing small, pea-like flowers above ground and then, after pollination has taken place, 'burying' the seed pods in the ground to ripen, hence its common name. The seeds are harvested for their oil and protein. The oil is used domestically for food and cooking and it is also converted into margarine and soap. The nuts are often used in soups and curries in the Caribbean and as roasted peanuts and peanut butter they form an important part of the diet of people in North America and Europe. Roasted and salted nuts are often offered as an accompaniment to pre-meal drinks in hotels throughout the Caribbean. The plant has other uses as well. After the oil has been extracted the residue, called groundnut cake, forms a valuable cattle and poultry feed. The leaves and other plant remains left after the nuts have been dug up are also used as a manure.

23

Family   Myrtaceae

Guava (*Psidium guajava*)
Other names   Yellow Guava, Apple Guava

The guava tree is a native of tropical America and was well known to the Arawak Indians, the original inhabitants of the Caribbean islands. The tree is small to medium size and rarely attains a height of more than twenty feet. It produces small, white, solitary flowers. The guava fruit is pear-shaped, about two inches long and turns to a yellowish-orange colour when ripe. The flesh inside is pinkish with a strong, aromatic flavour and is full of small seeds. Guavas are eaten raw or stewed and served as a dessert. They may also be offered as a starter at breakfast. The fruit is sometimes used to make delicious jams, jellies and pastes and is also tinned and exported to temperate countries. In some islands the leaves and bark are used to make a kind of tea recommended for the treatment of diarrhoea.

Family   Moraceae

Jackfruit (*Artocarpus heterophyllus*)
Other names   Jak Fruit

The jackfruit originates from the Indo-Malaysian region, but it is now found in most parts of the tropics. It is a member of the fig family and is related to the breadfruit tree; the two fruits resemble each other in appearance. The tree produces separate male and female flowers, both of which are greenish in appearance and are often found growing on the same branch. The tree is easily recognised by its massive, oval, green-coloured fruits. These are easily confused with those of the breadfruit. Both have the same 'pimply' skin but, unlike breadfruits, jackfruits are attached close to the bark on short stalks. Jackfruits sometimes grow to a massive size and a single fruit weighing forty pounds is not unusual. The fruits can be eaten raw or cooked. The white, fleshy and somewhat gelatinous inside is full of seeds and these are sometimes roasted and eaten like chestnuts. In some countries in the Far East the woody part of the seed is boiled and crushed to obtain a yellow pigment used for dyeing cloth.

Family Rutaceae

Lime (*Citrus aurantifolia*)
Other names West Indian Lime, Key Lime

This small, spiny tree is a native of north-east India and Malaysia, but it is now grown widely in the tropics, including the Caribbean. The fruits vary in size, but most of those seen in the Caribbean islands are usually spherical, oblong or ovoid in shape. The fruit is grown for its juice and for the oil which is extracted from the rind. Most limes grown in the Caribbean taste highly acidic. The first lime trees to appear there were planted by the early European settlers and since that time the fruit has become an important part of Caribbean cookery. Lime-juice is an important ingredient in Creole sauces and numerous fish, vegetable and poultry dishes. It is also used to flavour desserts and it makes an ideal companion to many rum cocktails. Limes are grown commerically in some Caribbean islands and Montserrat and Dominica are famous for the preparation and export of raw or concentrated lime-juice for use in making lime cordial drinks.

Family   Guttiferae

Mammee Apple (*Mammea americana*)
Other names   Mammee, Mamey, St Domingo Apricot

The mammee apple tree is a large evergreen native to the Caribbean region. Its fruits were already part of the local diet when Columbus first arrived in the late fifteenth century. The fruit itself is spherical in shape with a rough, brown skin. The fleshy interior is apricot-coloured and contains a number of large, blackish-brown seeds. The fruit can be eaten raw but is also stewed and eaten as a dessert. Fruits that are not quite so ripe are sometimes used to make jams and jelly-like preserves. In some parts of the Caribbean, an aromatic liqueur (eau de creole) is distilled from the highly scented flowers. The tree's fine-grained timber is also used extensively for furniture and cabinet-making.

Family Anacardiaceae

Mango (*Mangifera indica*)

The mango tree is easily recognised by its dome-shaped crown of dull green leathery leaves growing on a short, thick trunk. The small, green flowers grow in inconspicuous clusters. The most common fruit shape is kidney-shaped, each fruit hanging down on a long stalk. When the fruit is ripe, its skin is pinkish or yellowish, lightly flecked with black and brown. The flesh is yellowish-orange and is succulent yet fibrous, especially where it attaches to the single large stone. Mango is common in breakfast fruit platters and in fruit salad desserts. In some islands it is cooked as mango pie, mango brown Betty and even mango mousse. In Cuba it is pulverised and blended with milk to make a delicious cool drink. Unripe mangoes are used to make a chutney eaten as a side-dish with hot curries. The leaves produce a yellow dye and in some islands they are used as cattle fodder. The bark of the tree provides a tannin and the wood is used in shipbuilding.

Family Sapotaceae

Naseberry (*Manilkara zapota*)
Other names Sapodilla, Marmalade Plum, Beef Apple, Nispero, Chiku

The other popular name for this fruit is the sapodilla. The tree is a native of Central America and the West Indies. It is quite slow growing. It has oval-shaped, leathery leaves, each with a dark green, shiny surface. The flowers are small and white. The fruit is round in shape, about four inches in diameter and with a reddish-brown skin. The pulp has small, black seeds scattered inside it and is translucent and fragrant with a sweet taste. The crushed seeds are drunk as a treatment for oliguria. The fleshy pulp is used to make sapodilla custard and ice-cream. Chicle gum is extracted from the sap of the trunk and is used as an ingredient in the manufacture of chewing gum. This is a particularly important industry in Belize. Each tree should be tapped once every six years, but farmers often break this rule with the result that many trees become weakened and die of disease. The timber is hard, very durable and is valuable commercially.

Family Myristicaceae

Nutmeg (*Myristica fragrans*)
Other names Nutmeg Tree

The nutmeg tree first reached the Caribbean in 1824. It was introduced to Grenada in 1843 when a merchant ship called in there on its way back to England from the East Indian spice islands. The ship had a small quantity of nutmeg trees on board and some of these were left behind on the island. This was the beginning of Grenada's nutmeg industry, an industry that now supplys nearly forty per cent of the world's annual crop. The tree grows to about forty feet in height and its inconspicuous flowers produce small, peach-shaped fruits, each with a nut inside. The nut is covered with a scarlet aril layer which, when dried, is known as the spice mace. The outer part of the fruit is fermented to form a brandy-type drink. The seed is the nutmeg of commerce. Most nutmegs are ground into powder or crushed for their oil, but the most perfect specimens are exported whole. Nutmeg even has medicinal uses. Some sufferers from strokes keep a piece of nutmeg in their mouth to 'prevent' further attacks. The visitor doesn't have to go to Grenada to see nutmegs growing. There are plenty of trees in Welchman Hall Gully on Barbados, but the visitor should take care. Nutmeg and mace are poisonous if taken in large quantities due to the presence of an aromatic oil called myristicin.

Family Malvaceae

Okra (*Hibiscus esculentus*)

Other names Gumbo, Lady's Fingers, Syrian Mallow, Gombo, Okro, Ochro, Achro, Bamie

Okra is another plant that was brought to the Caribbean during the days of the slave-trade. It belongs to the same family of plants as the gaudy hibiscus whose flowers visitors see growing in all the Caribbean islands. The plant produces medium-sized yellow flowers and these develop pod-like fruits, each about six inches long. The pods are harvested about three months after the initial planting. Each pod is almost oblong in shape and tapers to a point at one end. The interior is rather soft and is filled with a mass of glutinous seeds. Okra is used as a vegetable in many Caribbean dishes. It is also eaten combined with a peppery preparation of sweet corn in a dish called coo-coo. It can be served as a vegetable dish to accompany a meat course and is even used as a side-dish with various curries. In some parts of the Caribbean the fruits are referred to as 'lady's fingers' because of their resemblance to a human hand. The seeds of okra are often used as a substitute for coffee while the leaves and immature fruits have a history of being used as a poultice. Decoctions of the fruit pod have a variety of medicinal uses in the different Caribbean islands. These range from the treatment of eye complaints and catarrh to inflammation of the reproductive system.

Family Rutaceae

Ortanique

This is a member of the citrus fruit family of plants. The ortanique first appeared in Jamaica and probably occurred purely by chance. It is a hybrid of a cross between the sweet orange (*Citrus sinensis*) and the tangerine (or mandarin, *C. reticulata*). The name ortanique reflects the origins of the fruit: it is a composite of *or*ange-*tan*gerine-un*ique*. It is very juicy and has a pleasant flavour. Like many hybrids, the ortanique bears characteristics of both parental stocks. It resembles the sweet orange in size and juice content but, like its other ancestor the tangerine, possesses a thin, easily-peeled skin. It is this latter characteristic which has made the ortanique a popular fruit in recent times. The fruit is grown on a commercial scale in a number of Caribbean islands including Jamaica.

Family Passifloraceae

Passion Fruit (*Passiflora edulis*)
Other names Granadilla

The passion fruit plant is a native of South America. There are at least four hundred different species and many of these are distributed throughout the tropics as well as in some temperate countries. It is a vine-like plant, which climbs by means of long, green tendrils. The flower has a characteristic and unique shape and the arrangement of the floral parts is supposed to symbolise the crucifixion, hence its name. Some species produce small ball-shaped fruits and one species in particular, *Passiflora edulis* var. *edulis*, is grown commercially for its fruits. The fruits are purple when ripe and the flesh is usually eaten raw in fruit salads. The pulp can also be squashed and the juice made into a cool and refreshing drink or even used in the manufacture of ice-cream to impart the distinctive passion fruit flavour.

Family Caricaceae

Pawpaw (*Carica papaya*)
Other names Papaya, Papaw, Tree Melon

In his journal, Columbus wrote that the natives of the Caribbean islands were very strong and lived largely on a tree melon called the 'fruit of angels'. He was talking about the pawpaw tree. Although it is classified as a tree, the stem of the pawpaw is largely hollow and the plant lacks the wood characteristic of normal trees. The stem is unbranched with a crown of large, palmate leaves springing from the top. There are separate male, female and hermaphrodite trees but only the female and hermaphrodite trees bear fruit. The fruits vary in size, some attaining a weight of ten pounds or more. The flesh is yellowish-orange when ripe with small, dark seeds scattered amongst it. Pawpaw is often eaten as a breakfast dish with a piece of lime squeezed over it. Both the leaves and the fruit contain papain, a protein-digesting enzyme. In some islands the local inhabitants tenderise their meat before cooking by wrapping it in pawpaw leaves. The enzyme can be extracted and is used commercially as a meat tenderiser. In some islands the leaves are boiled and eaten as a vegetable.

Family Solanaceae

Peppers (*Capsicum* spp.)
Other names Sweet Peppers, Pimenta, Pimiento, Red Peppers, Green Peppers, Paprika, Chilli Peppers

Peppers are a common sight throughout the Caribbean, where they are an important ingredient of many West Indian recipes. The fruits grow on small bushes and they are produced both commercially and for home consumption. There are two main types of pepper, chilli peppers and sweet peppers. Chilli peppers are small, red or green fruits which are used to impart a hot, spicy taste to soups, stews, sauces and curries. Visitors to the Caribbean will often see the harvested chilli peppers spread out on mats to dry in the sun. Sweet peppers grow to a larger size. They are also coloured red or green. They have a much cooler taste and are eaten raw as part of salads. They can also be cooked as a vegetable in stews, or served by themselves, stuffed with finely chopped meat or other types of filling.

Family Leguminosae

Pigeon Pea (*Cajanus cajan*)
Other names Congo Peas, Congo Beans, Goongoo Peas

This plant forms a perennial shrub which grows to a height of about nine feet. It is capable of withstanding extremes of drought and its ability to 'fix' nitrogen in the soil makes it a useful plant to introduce into a crop rotation scheme. There is some doubt about whether it originates in Africa or India. It is certainly widely cultivated throughout the tropics and is a particularly important crop in the Caribbean region. The mature pods range in colour from light green to dark brown and a well-established plant is capable of producing good, seed-bearing pods for a number of years. The seeds vary in colour from grey to yellow and resemble small garden peas in shape and size. They are often used in soups and curries after first being dried and split. Rice and pigeon peas is a favourite dish in a number of Caribbean islands, particularly Trinidad and Jamaica. In the latter island the peas are sometimes cooked in the form of a dumpling.

Family Bromeliaceae

Pineapple (*Ananas comosus*)
Other names Pine

'None pleases my tastes as do's the pine' wrote George Washington in his diary when he visited Barbados in 1751. The pineapple takes its name from the resemblance it bears to the pine cone of temperate regions. It is a native of Central America and the West Indies and Columbus found it in this area when he arrived in 1492. Since then the plant has spread all over the world and is now a popular fruit in all tropical countries. Because of the structure and arrangement of its leaves, it can make maximum use of rain-water and is capable of living in very dry conditions. It flourishes in most Caribbean islands and is the largest export crop of Puerto Rico. The flowers arise deep in the centre of the leaves where they bloom for only a brief period. About ten months later, the pineapple has grown and ripened and is ready for picking. It is used widely as a breakfast dish, a dessert and in fruit salads. It also accompanies meat dishes such as gammon and pork and is even used to make a delicious squash drink. A particularly delicious variety is the 'Antigua Black'. Forms of the plant with variegated leaves are often grown as indoor house-plants.

41

Family Musaceae

Plantain (*Musa* sp.)
Other names Green Banana

Plantain is the name given to the green form of the banana. Plantains do not ripen in the same way as their cousins the sweet bananas and they are always less sweet. The individual fruits are often bigger than a sweet banana and each is usually more clearly horn-shaped. Plantains are one of the most prolific of all carbohydrate producing crops and it has been calculated that an area of ground which is capable of producing fifty pounds of wheat or one hundred pounds of potatoes could carry as much as four thousand pounds of plantains. Although plantains are less sweet than sweet bananas, they are far more versatile in their use. They are often boiled and served with meat as part of a main course but because their tissue has a 'starchier' taste than sweet bananas they are best cooked with plenty of spices, onions and pepper, or mixed into a hot casserole. In some islands they are cut into thin slices and fried in a pan of deep fat to produce delicious crisps. Chopped beef and plantains are the main ingredients of the famous Puerto Rican dish *piononos*. In this dish the plantains are sliced wafer-thin and then fried and wrapped around a mixture of cheese and finely-chopped meat.

Family Punicaceae

Pomegranate (*Punica granatum*)
Other names Apple of Carthage

Visitors to the Caribbean will come across pomegranates from time to time. They are certainly to be found on the British Virgin Islands and also on Barbados. The fruits are borne on a medium-sized tree about twenty feet in height with characteristically spiny branches. The flowers are reddish, sometimes even scarlet in colour, and they have crinkly, paper-like petals. The fruit is round in shape and about the size of an orange. It has a thick, smooth, leathery skin which turns yellowish-red when the fruit is ripe. The pulp inside is very juicy and contains a mass of small, white seeds. It is usually eaten raw but sometimes the juice is used to make a drink called grenadine. The peel of the fruit is rich in tannins which can be used to produce high quality leather.

43

Family   Cucurbitaceae

Pumpkin (*Cucurbita* sp.)
Other names   Cushaw Pumpkins, Hundredweight Fruits, Winter Squash

Pumpkins are large, spherical fruits produced by vine-like plants belonging to the gourd family. The mature fruit has a brownish-yellow skin with a slightly ribbed surface. Pumpkins are grown widely throughout the tropics and they are particularly popular in many Caribbean islands. The seeds are sown in deep holes filled with a rich mixture of soil and manure and, after germination, the vines are allowed to grow over the ground with no support. The large fruits are commonly seen in markets and at roadside stalls. Sometimes they are sold whole but, because of their large size, some vendors cut them up and sell smaller portions. Pumpkins are one of the largest fruits known and, for this reason, they are sometimes called 'hundredweight gourds'. Individual specimens often exceed six feet in circumference and the biggest pumpkin on record weighed in at a massive 493½ pounds!

44

Family Gramineae

Rice (*Oryza sativa*)
Other names Paddy

The rice plant is a member of the grass family and is one of the world's most important crops, since it forms the staple diet of more than one third of the human population. It is a unique plant and certainly one of the few grasses that requires a flooded soil in which to grow. Flooded paddy-fields are a common sight in many parts of the Far East but they are less so in the Caribbean. Rice needs low-lying swampy land to grow properly and Guyana has the ideal conditions to make rice a successful commercial venture; it is the main rice producer of the Caribbean islands. The rice grains we eat are the fruits of the plant after the female flowers have been fertilised and their ovaries have become swollen and finally mature. Rice itself is rich in carbohydrates but has few vitamins. The husk surrounding each grain is rich in B vitamins but this is often removed during the processing of the grains. 'Beans and rice' is a common dish throughout the Caribbean and each main island has its own version and recipe. In Jamaica the dish is called rice and peas but the 'peas' are either red kidney beans, small red beans native to the island or pigeon peas. The words 'peas' and 'beans' are interchangeable throughout the islands.

Family Polygonaceae

Sea Grape (*Coccoloba uvifera*)

Some historians have suggested that the sea grape was the first plant Columbus saw when he came to the Caribbean in the fifteenth century. It is certainly characteristic of many Caribbean beaches and visitors will soon be familiar with its smooth, leathery leaves with their prominent reddish veins. The male and female flowers occur on separate plants and after fertilisation, the fruits develop into purple, grape-like clusters. The clusters of fruits do not ripen evenly. The sea grape varies in habit depending upon the environment in which it is growing. On exposed shores it grows as a sprawling shrub but in more sheltered areas it grows as a tree, reaching a height of about fifty feet. The sour fruits are edible when ripe but are too sour to be really pleasant. They contain large amounts of pectin which makes them suitable for making good jelly and jam preserves. In some islands they are made into soup.

Family   Rutaceae

Shaddock (*Citrus grandis*)
Other names   Pummelo, Pamplemousse, Pompelmoose

The shaddock is a type of citrus fruit which originates from the islands of Polynesia in the South Pacific. It was brought to the island of Jamaica in the seventeenth century by the captain of a British merchant ship who gave his name to the fruit. The shaddock is a large fruit about the size of its more modern relative the grapefruit. It has a characteristically thick skin or rind and a yellowish-red, fleshy pulp which has an aromatic scent and a rather bitter taste. Since the time of its introduction, another form of citrus fruit has evolved from the shaddock, either naturally through genetic mutations, or as a result of man's experiments with cross-breeding and hybridisation. This new arrival is called the grapefruit and you can read more about it on page 21.

Family   Malvaceae

Sorrel (*Hibiscus sabdariffa*)
Other names   Roselle, Jamaican Sorrel, Indian Sorrel, Red Sorrel

Sorrel is a small, annual plant which grows to a height of about six feet. After flowering, the petals of the individual blooms wither and drop off leaving behind the protective sepals. The sepals gradually swell and become fleshy and succulent. Eventually the sepals envelop the seed pod formed from the female parts of the flower and it is these large, swollen sepals which are used as the 'fruit'. They are usually ready for picking just before Christmas and they are commonly used in preparation for the Christmas festivities. The mature sepals are mixed with other flavourings and soaked in water for a few days to make a dark red, aromatic drink with rather a sharp taste. In some islands, including Trinidad, the fruits are gently fermented and then added to rum to produce a very potent liqueur. The ripe seeds are also used to make excellent jams and jelly preserves.

Family Annonaceae

Soursop (*Annona muricata*)
Other names Guanabana

The soursop tree is a native of tropical America and is found in many of the Caribbean islands, where its fruit is particularly popular. It is a fast-growing tree with small, shiny leaves and yellowish-green flowers. The fruit is a large, oval-shaped structure which is dark green in colour. Its skin has a distinct spiny surface and is easily recognised in markets and on roadside stalls. Some fruits attain a length of eight inches, and a weight of six pounds is not unusual. The inside of the fruit is whitish-pink and the flesh has the consistency of thick custard. The fruit is seldom cooked. More frequently it is sieved and served as cream or soursop ice-cream. It can also be made into drinks and sherbert. It retains its flavour even after deep-freezing for a period of time.

49

Family   Sapotaceae

Star Apple (*Chrysophyllum cainito*)

The star apple tree grows to about fifty feet in height and is often planted as a shade tree in many Caribbean islands because of its dense foliage, which is shed only very infrequently. The tree produces small, inconspicuous purplish-white flowers. The mature fruits are about the size of an ordinary apple and they have a dark, smooth, purplish skin when they are ripe. The fruit gets its name from the pattern revealed when it is cut in half. The star shape is produced by the arrangement of the purplish seeds amongst the white, rather gelatinous flesh. The fruit has a sweetish but mild taste. The tree is quite sensitive to cold conditions and although it grows throughout the Caribbean, it is more common in Jamaica and Haiti. The fruit is eaten raw or as a dessert, either as part of a fruit salad or in a special mixture with pieces of orange.

Family Gramineae

Sugar Cane (*Saccharum officinarum*)

Sugar cane is a giant representative of the grass family. It has a leafy stem which is renewed each year from an underground rhizome. The flowers appear as tall, feathery plumes resembling the flower heads of pampas grass, more familiar to gardeners from temperate regions. The visitor will see sugar cane growing on a number of the islands, but Trinidad, Guyana, Cuba and Barbados are particularly important producers. Cuba is the world's largest exporter of sugar and the crop accounts for over seventy per cent of Cuba's export earnings. In the days of the slave-trade, large numbers of men and women were needed to work the plantations. One historian has recorded that to run even the smallest plantation required 250 slaves! The sugar is extracted from the mature crop by crushing the stems. Apart from sugar, by-products of the industry include molasses from which rum is made. The different islands have their own characteristic rum. You need ten tons of sugar cane to make thirty gallons of rum!

Family   Gramineae

Sweet Corn (*Zea mays*)
Other names   Indian Corn, Corn, Maize

Sweet corn, or maize as it is commonly known, is grown in many Caribbean islands although, as a food source, it is probably not as important as it is in many African countries. The plant grows to a height of about seven feet. It produces separate male and female flowers; the male flowers are carried at the top of the stem and the female flowers about half-way down. The male pollen is carried to the female flowers by wind. Each female part is really a collection of many small flowers arranged around a central axis. Each small flower is capable of being fertilised and then developing into a fruit or maize grain. When fully matured the swollen fruits are called cobs and it is these which are picked and used for food. The cobs can be roasted, baked or boiled. They can be served as a starter to a main course or given as a vegetable to accompany other dishes. A number of by-products are obtained from the grains including corn oil used in the manufacture of soap, as a cooking oil and in salad dressings. One of the best known uses of corn is in the breakfast cereal 'Cornflakes'.

Family   Convolvulaceae

Sweet Potato (*Ipomoea batatas*)
Other names   Sweet Potato Vine

The sweet potato is an important vegetable and forms the staple diet of many people living in the tropics. The plant is a leafy vine which develops large swollen roots beneath the surface of the soil. These are root tubers and are different from the tubers of the Irish potato, which are derived from the plant stem. When mature, the sweet potato forms an elongated oval-shaped structure with a smooth, reddish skin. The tuber is rich in carbohydrate, particularly starch. It is usually eaten boiled or roasted. In some islands it is called 'yam' but this is misleading. The genuine yam comes from a different plant altogether. Sometimes the tubers are preserved by canning or dehydration and used as a source of flour, starch, glucose, syrup and alcohol. The green remains are used as an animal feed.

Family Leguminosae

Tamarind (*Tamarindus indica*)

The tamarind tree grows to a height of about fifty feet. It is a tall, graceful tree and is often planted as a wind-break in many Caribbean islands, where it gives protection against hurricanes and strong winds. The ripe fruit forms a small brown pod about four inches long and when mature the pod has a pulpy interior surrounding several large seeds. When still green, the fruit is slightly acidic and is often used as a seasoning for fish and meat dishes. In the unripe state it is also added to curries and similar dishes. The ripe fruit is sometimes used to make a sugary sweet or candy which has a very spicy taste. The fruit can also be crushed to make a drink with a strong aromatic flavour and refreshing quality. Tamarinds are perhaps most famous for the part they play in the processing of Angostura Bitters. In Barbados the pulp of the fruit is used to make tamarind balls which are to be found in most supermarkets.

Ugli (*Citrus* sp.)
Other names Hoogly

This unusual fruit has come about as the result of various plant breeding and hybridisation experiments which have taken place in the Caribbean over many years. The fruit gets its strange name from its unfortunate and misshapen appearance. The people of Jamaica call it 'hoogly'. The fruit is really a cross between a grapefruit and an orange. Although its skin is thick and lumpy and certainly looks unattractive, the fleshy fruit within is considerably sweeter than most grapefruits and usually contains very few seeds. The fruit is grown commercially in Jamaica. Other crosses between different citrus fruits have been produced. Among the hybrids recently developed are the tangelo, a cross between a tangerine and a grapefruit and the tangor, which has resulted from interbreeding between a tangerine and an orange.

Family Cucurbitaceae

Watermelon (*Citrullus lanatus*)

Watermelons are a common sight in most Caribbean islands and the visitor often sees them for sale at the roadside when travelling from the airport to his hotel shortly after arrival. The plant itself has a creeping habit with large, dull green leaves. The flowers are usually white or yellowish in colour. The fruits often grow to a large size and some specimens may reach twenty pounds in weight. The skin is smooth and sometimes has a pattern of darker lines. The flesh inside is white, yellowish or deep pink with a large number of small, white, red or black seeds. The water content is very high and a plate of watermelon makes a thirst-quenching and refreshing meal. It is usually served as a breakfast starter or chopped up as part of a fruit salad. It can also be served as a dessert by itself at the end of a main course. The seeds contain large amounts of oil which is sometimes extracted and used for cooking and for fuel in oil lamps.

Family   Dioscoreaceae

Yam (*Dioscorea* sp.)
Other names   Cush-Cush Yam

The yam is a native plant of tropical America and was certainly an important part of the diet of many Caribbean people long before the arrival of Columbus and the Spanish explorers who followed him. The plant grows as a vine, sometimes reaching a height of six or eight feet. The edible part is formed from the swollen root tubers. These vary in size and large specimens may weigh as much as twenty pounds. The swollen tubers have a thick, rough skin which is usually dark brown in colour and hairy to the touch. Inside is a dense white and sometimes reddish fibrous material which is rich in carbohydrate, particularly starch. It is harvested in the dry season and will keep for several months stored in a dark, airy room. Yam still forms the staple diet of a large number of people in the Caribbean. It is cooked in different ways including boiled, baked or fried. It is served as a vegetable with meat although it can also be pounded into a meal. Some species of yam are sometimes called 'drug yams' because they yield a drug called diosgenin, used to produce oral contraceptive drugs.

Natural History and Related Titles

Bermuda's Botanical Wonderland   Phillips-Watlington
Birds of the Eastern Caribbean   Evans
Birds of Trinidad and Tobago   ffrench
Butterflies and Other Insects of the Eastern Caribbean   Stiling
Butterflies of the Caribbean and Florida   Stiling
Caribbean Wild Plants and Their Uses   Honychurch
Container Gardening for the Caribbean and the Tropics   Light
Coral Reefs of the Caribbean, The Bahamas and Florida   Lee and Dooley
Fauna of the Caribbean: The Last Survivors   Sutty
Fishes of the Caribbean Reefs   Took
Flowers of the Caribbean   Lennox and Seddon
Fruits and Vegetables of the Caribbean   Bourne, Lennox and Seddon
Gardening in the Caribbean   Bannochie and Light
Gardens of the Caribbean   Collett and Bowe
Growing Orchids in the Caribbean   Light
Marine Life of the Caribbean   Jones and Sefton
Native Orchids of the Eastern Caribbean   Kenny
Seashells of the Caribbean   Sutty
Seashell Treasures of the Caribbean   Sutty
The Ephemeral Islands: A Natural History of the Bahamas   Campbell
Trees of the Caribbean   Seddon and Lennox
Wild Plants of Barbados   Carrington
Wild Plants of the Eastern Caribbean   Carrington